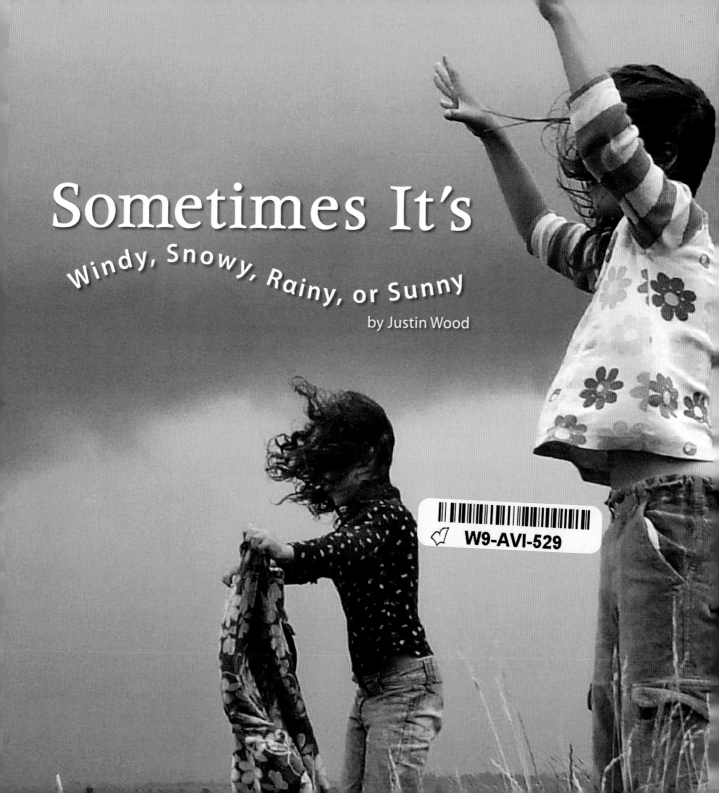

Sometimes It's

Windy, Snowy, Rainy, or Sunny

by Justin Wood

Fall

Sometimes it's windy in fall.
Kites fly and leaves dance.

2

What can you do on a windy fall day?

Winter

Sometimes it's snowy in winter.
The temperature of the air is cold.

What can you do on a snowy winter day?

Spring

Sometimes it's rainy in spring.
Rain falls and puddles form.

What can you do on a rainy spring day?

Summer

Sometimes it's sunny in summer.
The sun's energy warms the air.

What can you do on a sunny summer day?

 # Fall

 # Winter

What can you do in each season?

Spring

Summer